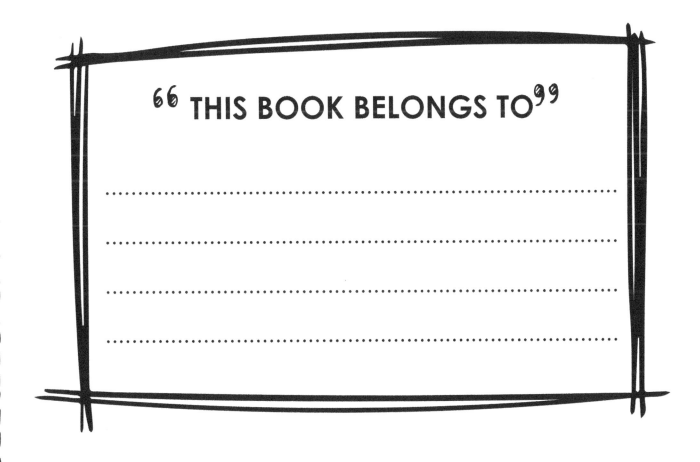

"THIS BOOK BELONGS TO"

....

....

....

....

COLOR TEST PAGE

COLOR: COLOR: COLOR: COLOR:

COLOR: COLOR: COLOR: COLOR:

COLOR: COLOR: COLOR: COLOR:

COLOR: COLOR: COLOR: COLOR:

COLOR: COLOR: COLOR: COLOR:

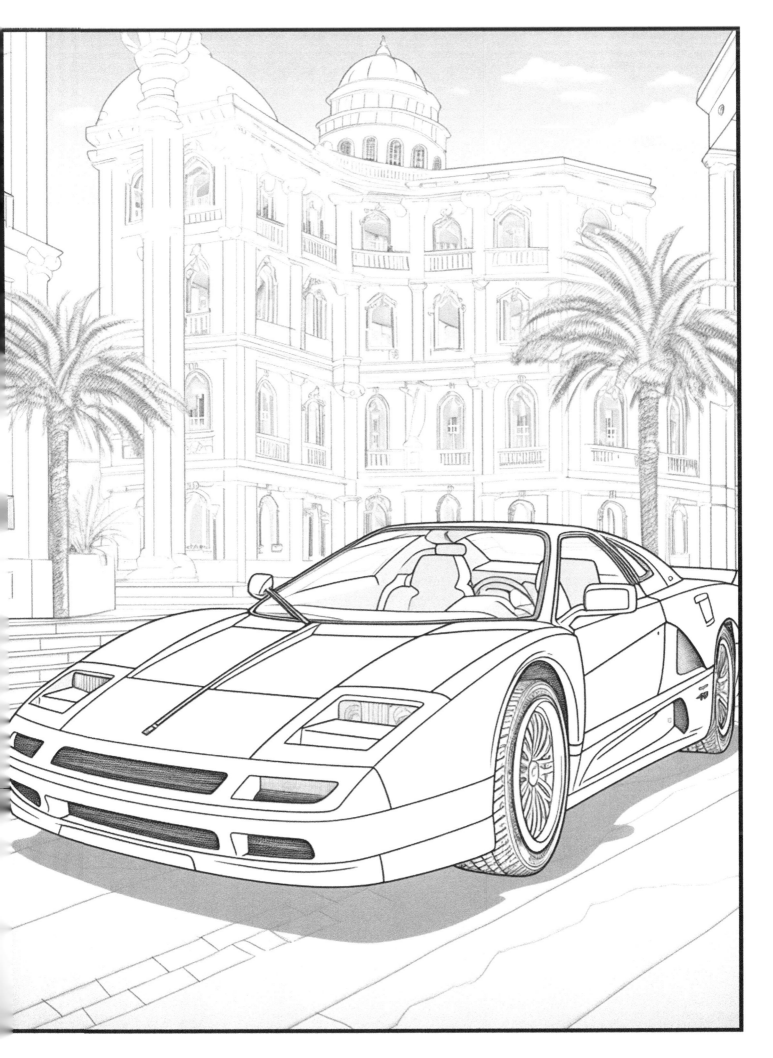

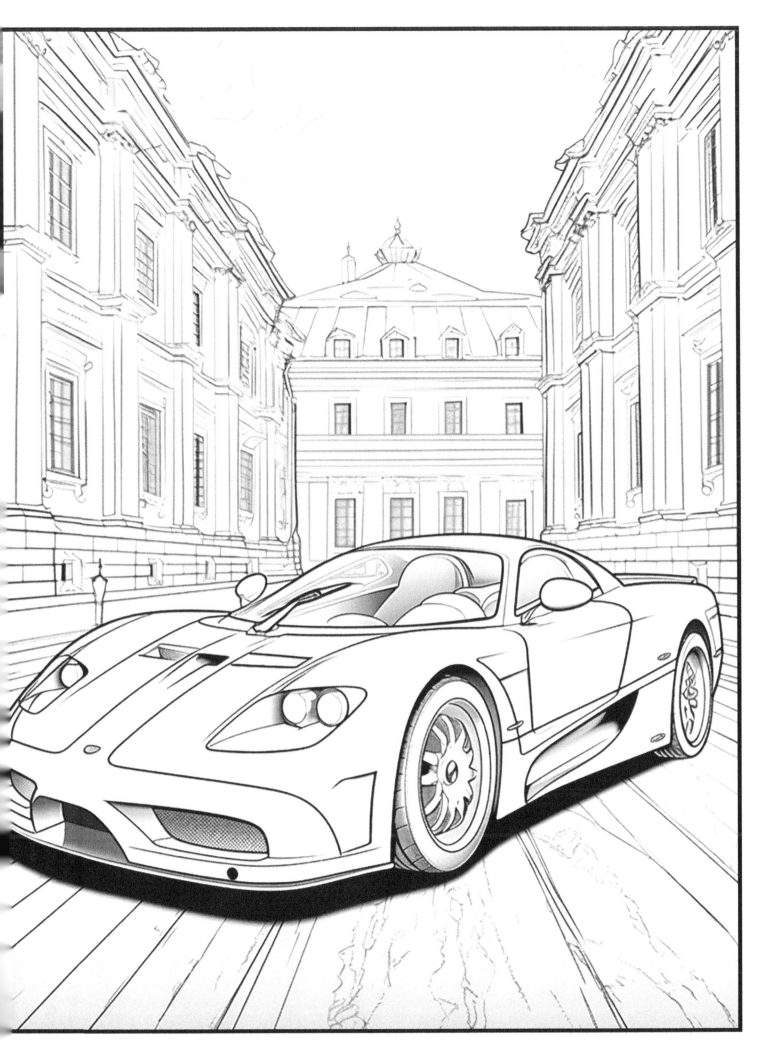

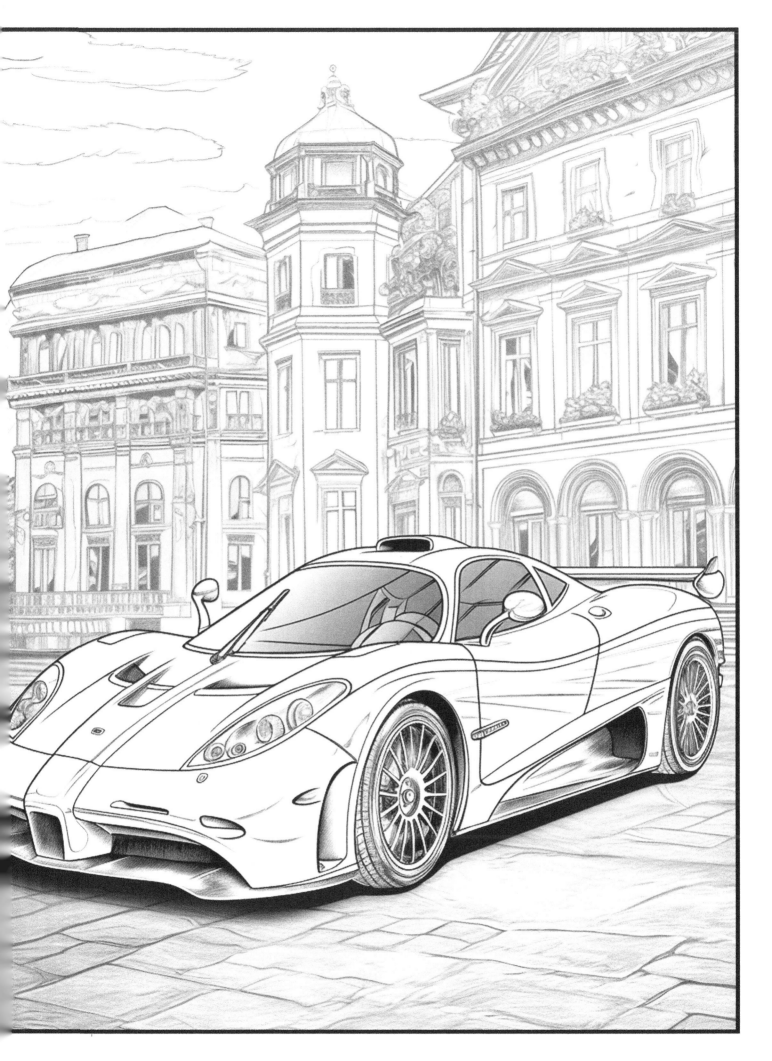

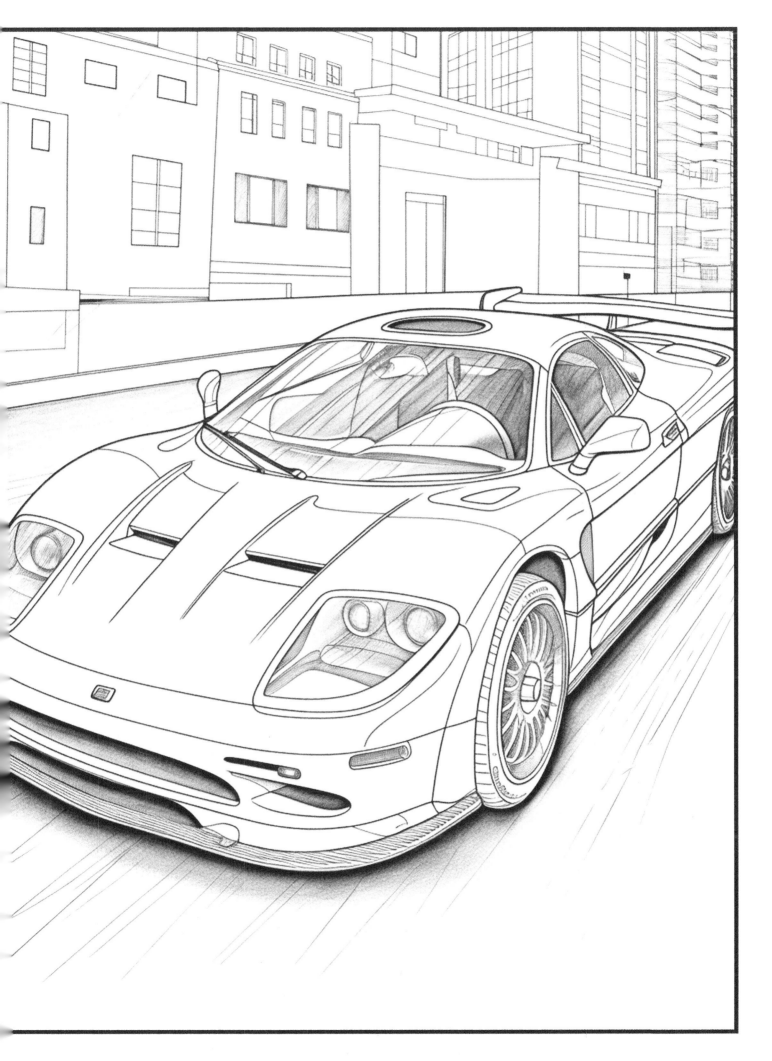

WRITE DOWN YOUR FAVORITE ASPECTS OF THIS BOOK:

THANK YOU FOR TRUSTING US BY PURCHASING OUR BOOKS

Your trust in us means a lot, and we truly hope that you will find joy and satisfaction in coloring our unique designs. If our book meets your expectations, we kindly ask you to leave a positive review as it motivates us to create even better books in the future. Once again, thank you for your support and we hope that our coloring book will bring a little bit of creativity and relaxation into your life.

Made in the USA
Monee, IL
05 December 2023

48221510R00044